UX
For Cats

50+ **Illustrated** Fundamental UX Terms & Facts for Beginners and Business Users **So Simple** Your **Cat** Can Understand Them

Written & Illustrated by
Douglas Bonneville

GotoCats™ Press

Dedicated to:

*Gucci, Rosie, Baghira, Trippy, Simba,
Max, KittyWac, Daisy, Humphrey, Cricket,
2nd Max, Fatty Cakes, and Coco*

ISBN: 978-0-9600439-2-7

Published in 2024 by GotoCats™ Press

Warwick, Rhode Island 02888 USA

For more information about books published by GotoCats™ Press, visit:

https://gotocats.com

Table of Contents

Introduction 9

Foundational Concepts 11

User Experience 12

User Interface Design Principles 13

User-Centered Design 14

Design Thinking 15

Information Architecture 16

Research & User Analysis 19

User Persona 20

User Journey 21

User Flow 22

User Feedback 23

User Interviews 24

User Surveys 25

User Stories 26

Task Analysis 27

User Research 28

Empathy 29

Design & Prototyping Processes 31

Prototyping 32

Wireframing 33

Rapid Prototyping 34

Iterative Design 35

Lean UX 36

Table of Contents

Visual & Interaction Design   39

Visual Design   40

Typography   41

Color Theory   42

Visual Hierarchy   43

Fidelity   44

Interaction Design (IxD)   45

Development Methodologies   47

Agile Development   48

Design System   49

Mobile First Design   50

Responsive Design   51

Inclusive Design   52

Usability & Accessibility   55

Usability Testing   56

Usability Guidelines   57

Cognitive Load   58

Accessibility   59

Error Prevention   60

Strategic & Analytical Approaches   63

Content Strategy   64

Data-driven Design   65

Ethical Design   66

Gamification   67

Table of Contents

Key Performance Indicators   68

A/B Testing   69

Conversion Rate   70

Click-through Rate (CTR)   71

Competitor Analysis   72

Additional Concepts   **75**

Consistency   76

Navigation   77

Above/below the fold   78

Information Visualization   79

User Engagement   80

Questions & Answers   **83**

Illustration & Memorization   **93**

Gallery   **93**

What to   **113**

Read Next   **113**

Beginner   **114**

Intermediate   **115**

Index   **117**

Feedback   **121**

Other Titles   **122**

About the Artist & Author   **123**

Introduction

Welcome to *UX for Cats*! If you've gotten this far, you're definitely intrigued. As they say, "curiosity killed the cat," but now we've caught you in our claws! Let's proceed!

UX for Cats is grounded in the scientific principle that our brains encode memories more effectively when the "cognitive load" is distributed across multiple modalities. In other words, when information is conveyed through text, images, sound, smells, or other cues, lasting memories are formed more easily. For this book, we utilize "dual coding" by presenting new information using text and illustrations. Our unique twist? We add humor (specifically, cat humor) as an additional factor to further embed the memories into your synapses!

The text for **each term is limited to three sentences,** and the images are drawn in an **easy-to-understand, simplified cartoon style** with enough visual richness to leave a fun impression. Reading the brief text while relating it to the image, any way you can, immediately creates stronger neural pathways than the short text could on its own. **The humor or novelty of each image further enhances those pathways.** Considering the book covers 50 fundamental definitions for the most essential keywords in UX, we offer a potent, unique, information-rich, and, above all, entertaining introduction to UX. This will greatly benefit you as you continue your studies with more in-depth books and video content.

Thank you for picking up this book! You're in for an exciting ride that promises to be as informative as it is enjoyable. We're here to prove that learning doesn't have to be tedious—it can be something you eagerly anticipate. As you flip through these pages, remember that each term and its accompanying cat-themed illustration are designed not just to teach, but to make a lasting impression.

Here's to diving deep into UX with all four paws, grasping every crucial term, and enjoying every moment of it!

Foundational Concepts

User Experience

User Experience, or UX, pertains to the entire experience a user has while interacting with a product or service, usually in relation to **how easy or pleasurable it is to use**. It includes aspects such as design, layout, text, and interaction elements. The concept of UX is end to end, reflecting a user's overall perception of a product's usability and ultimate worth.

User Interface Design Principles

User Interface Design Principles are rules and guidelines, as part of an overall process, that designers follow to create user interfaces (UI) in software or computerized devices, with a focus on usability, aesthetics, and style. These principles guide the design of digital assets like websites and apps, ensuring they are **intuitive and user-friendly**. They provide guidance on managing user's actions, also including how to display information, handling different screen sizes, and accessibility.

User-Centered Design

User-Centered Design refers to a design philosophy and process that places the needs, wants, desires, and even limitations of end users of a product at the forefront during each stage of the design process. This approach involves **testing and refining the product** over many cycles based on user feedback. The goal of User-Centered Design is to ensure that the final product is tailor-made to suit user needs.

Design Thinking

Design Thinking is an innovative, user-centric approach to problem-solving that involves understanding user needs, challenging assumptions, and redefining problems to identify alternative solutions not instantly apparent. This **iterative process** involves empathy, definition, ideation, prototyping, and testing stages, fostering a deep understanding of the user and the problem at hand. It's a vital strategy in the business environment to drive innovative solutions and competitive advantage.

Information Architecture

Information Architecture (IA) is the science of organizing and structuring information in products and services to support usability and findability. It involves creating a **structured layout of information** and defining how users interact with and navigate through a product. Good information architecture makes it easier for users to understand where they are in a system and where the information they want is located.

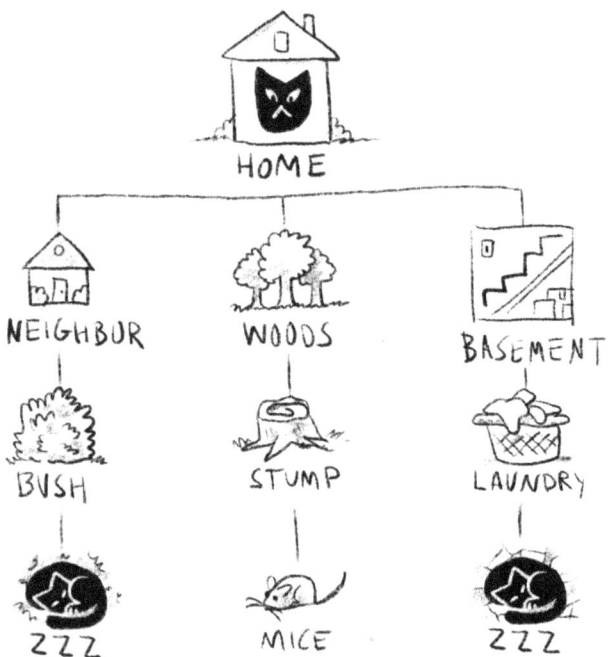

User experience (UX) is a term coined by Don Norman, a cognitive scientist and co-founder of the Nielsen Norman Group. He first used it in the late 1990s, when he was a vice president of advanced technology at Apple. He wanted to encompass all aspects of the user's interaction with a product, not just the usability or the interface.

User Persona

A User Persona is a fictional character created to represent a user type that will use a site, brand, or product in a particular way. This character is based on research and includes details like **demographics, behavior patterns, motivations, and goals**. Designers use personas to try to predict how potential users will interact with a product, and tailor their design solutions for them.

User Journey

A User Journey is a series of steps that represent a user's interactions with a product. It visualizes a user's path through a product, **from initial contact through the process of engagement and into a long-term relationship**. Understanding the user journey can help designers create a product that meets user needs and expectations.

User Flow

User Flow is a visual representation of the path a user follows through an interface to achieve a goal. It includes **all the different paths and touchpoints a user interacts with**. Understanding the user flow can help designers identify potential issues and improve the overall user experience.

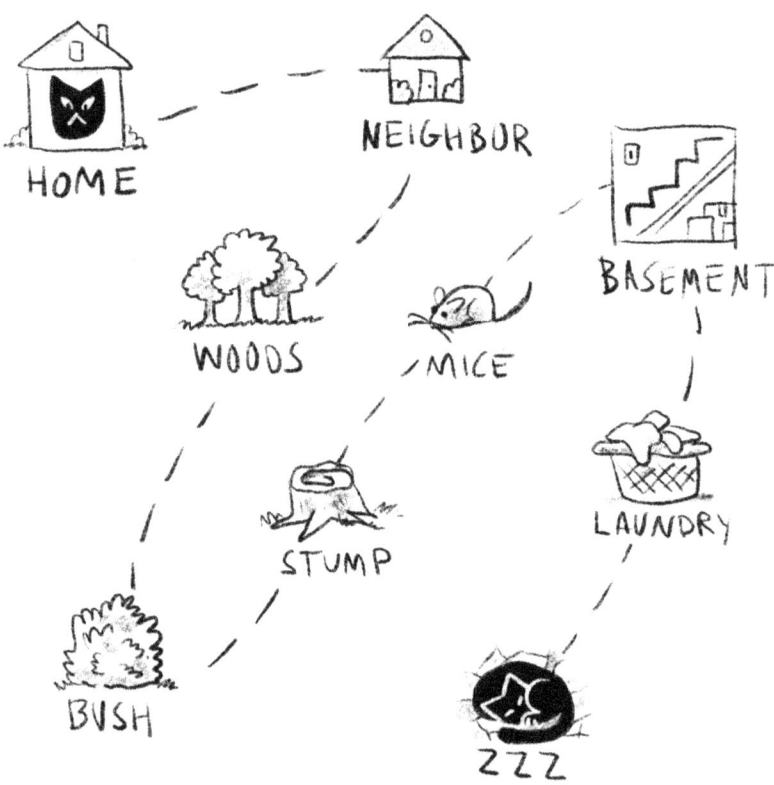

User Feedback

User Feedback is information provided by users about their experience with a product or service. It can be collected through **surveys, interviews, usability testing, and more**. This feedback is crucial for understanding user needs, improving products, and enhancing the overall user experience.

User Interviews

User Interviews are a research method used in design to gain insights into users' needs, behaviors, experiences, and motivations. They involve a structured conversation with users, where they are **asked about their experiences and perspectives on the product or service**. User Interviews help designers to understand their audience and create more effective designs.

User Surveys

User Surveys are a research tool used to gather feedback from users about a product or service. They are usually made up of **a series of questions and can be conducted online, in-person, or over the phone**. Surveys help to collect user opinions, understand user behavior, and improve the overall product based on the feedback received.

User Stories

User Stories are a tool used in Agile software development to capture a description of a software feature from an end-user perspective. They help the team understand the value of the feature and how it will be used. User Stories are typically written in the format: **'As a [type of user], I want [an action] so that [a benefit or value]'**.

Task Analysis

Task Analysis is the process of breaking down a task into its smaller steps to understand how users perform it. It's used to **identify potential problems, understand user behavior, and design more effective and user-friendly products**. Task Analysis can lead to more intuitive interfaces and improved user satisfaction.

User Research

User Research is the process of understanding user behaviors, needs, and motivations through observation techniques, task analysis, and other feedback methodologies. This understanding can help to create products that are **more aligned with user needs and expectations**. User Research is a crucial part of the design process that can lead to better products and higher user satisfaction.

Empathy

Empathy in design is the practice of understanding and sharing the feelings of users. It involves **putting yourself in the user's shoes to understand their experiences, needs, and perspectives**. Designing with empathy can lead to more inclusive, user-friendly products that meet the needs of a wide range of users.

Alan Cooper, known as the "Father of Visual Basic", invented the concept of user personas in the 1990s. He created user personas based on his observations and interviews with real people who used software. He used user personas to guide his design decisions and communicate them to his clients.

Design &
Prototyping
Processes

Prototyping

Prototyping is the creation of a preliminary model or release of a product to test its functionality, design, and usability. It's a critical stage in the design process, allowing for **validation of ideas, identification of design flaws**, and iterative improvements. In a business context, prototyping minimizes development costs and risks by allowing potential flaws to be addressed before the final product is built.

Wireframing

Wireframing is a process to create a simple sketch or layout of a webpage or app to illustrate the basic structure and components of a page. It's used in the early stages of design to establish the basic structure of a page **before visual design and content is added**. It helps designers and clients visualize the layout, navigation, and overall function of a site or app without having to worry about a specific design or style too early in the process.

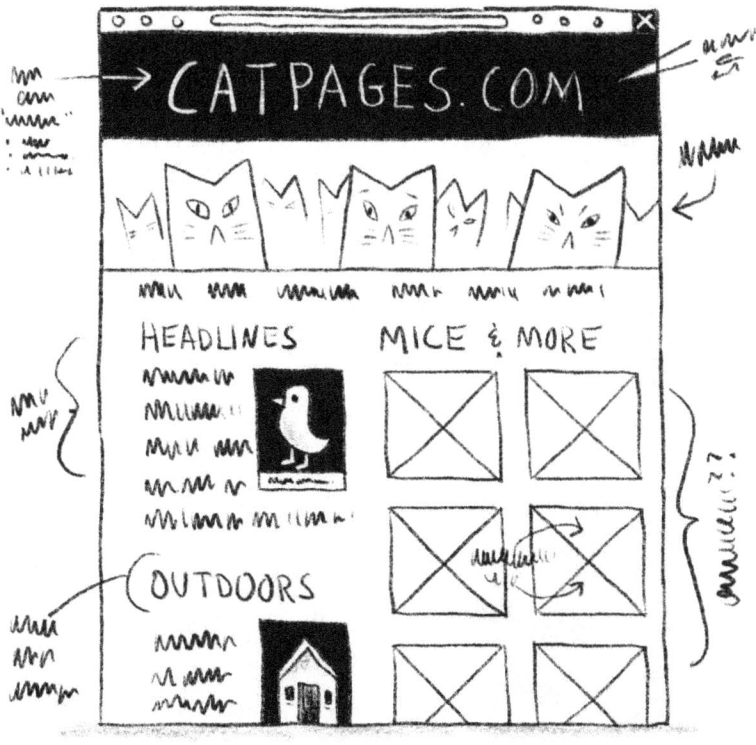

Rapid Prototyping

Rapid Prototyping is a design process where prototypes are quickly created and tested for functionality, performance, and feedback. It's a cycle of designing, prototyping, and testing, allowing for **quick iterations based on feedback**. Rapid Prototyping can save time and resources and lead to a better final product.

Iterative Design

Iterative Design is a design approach that involves a cyclical process of prototyping, testing, analyzing, and refining a product, with the aim of learning and enhancing the product through each cycle. This method allows for the **incorporation of user feedback into the design process continuously** as it comes up. Iterative Design can result in a better product fit, reduced development costs, and increased customer satisfaction by allowing continuous improvement and alignment with customer needs.

Lean UX

Lean UX is a design philosophy that seeks to streamline the UX design process by minimizing extensive documentation and maximizing focus on creating the user experience. This methodology encourages **quick, iterative design cycles**, emphasizing user feedback and making adjustments accordingly. Lean UX facilitates faster time to market, reduces waste, and ensures a stronger alignment with user needs than other methods that value process over the customer.

The first prototype of the computer mouse was made of wood and was designed by Doug Engelbart in 1964. It had two gear-wheels perpendicular to each other, which could detect movement in the X and Y directions. The mouse was later refined and popularized by Apple.

Visual & Interaction Design

Visual Design

Visual Design refers to the aesthetics of a product and how it improves user experience by organizing all the elements and data in a design in a way that is **pleasing to the eye**. It involves the use of colors, shapes, images, typography, and form to enhance usability and improve user experience. A well-designed product can guide users through a system and influence their feelings towards the product.

Typography

Typography is the art and technique of arranging type to make written language legible, understandable, and appealing when displayed. It involves the selection of **typefaces, point size, line length, line-spacing, letter-spacing and other factors**. Good typography can influence the reader's mood, cognitive load, and perception of the content in a positive way that leads to a good user experience.

Color Theory

Color Theory is a set of principles used by designers to choose, combine, and apply colors. It's a vital part of visual design, helping to communicate messages, evoke emotions, and **guide users through an interface**. Understanding color theory can help designers create visually pleasing and effective designs while also being inclusive of users who may have disabilities.

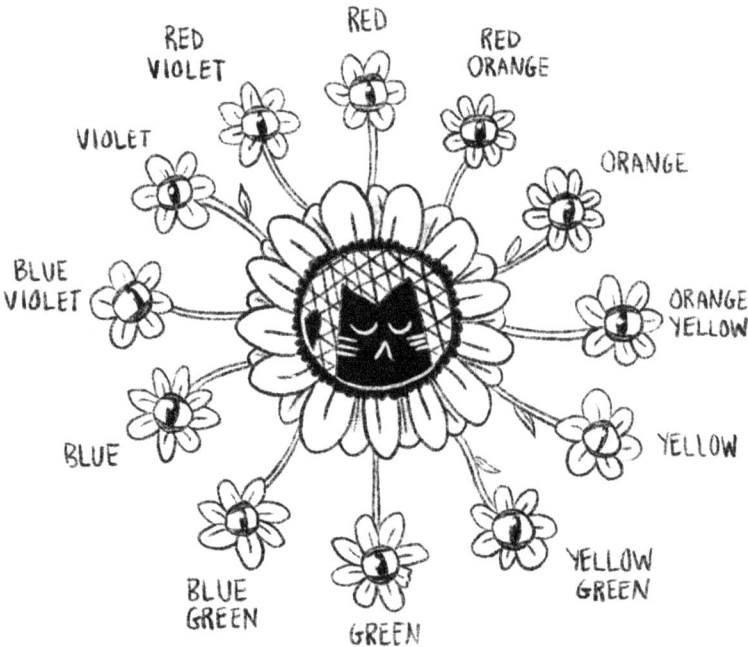

Visual Hierarchy

Visual Hierarchy in design refers to the arrangement or presentation of elements in a way that implies importance. It guides the eye to move in **a certain order or pattern**, influencing the order in which information is perceived. Visual Hierarchy can make designs more understandable and can guide users to take desired actions.

Fidelity

In design, fidelity refers to the level of detail and functionality in a design representation, such as a prototype. Low-fidelity designs are **more basic** and focus on functionality and structure, while high-fidelity designs are **more detailed** and closely resemble the final product. Different levels of fidelity can be used at different stages of the design process to test and improve the design.

Interaction Design (IxD)

Interaction Design, or IxD, is the design of the interaction between users and products. It involves creating **intuitive and efficient interfaces** that enable users to achieve their goals. Good Interaction Design can make a product easier to use, more enjoyable, and can lead to higher user satisfaction.

The term "graphic design" was first coined in 1922 by William Addison Dwiggins. He used it to describe his work in book design and typography. Dwiggins is thus credited with naming the profession.

Agile Development

Agile Development is a methodology for software development that emphasizes flexibility, collaboration, customer satisfaction, and rapid delivery that demonstrates value as early in the process as possible. It involves **regular reflection and adjustment in response to changes** and customer feedback through the design and development processes. This approach allows for a more flexible and adaptive development process as opposed to traditional "waterfall" approaches that make dealing with changes much harder.

Design System

A Design System is a collection of standards, components, and principles that are used to guide the creation and iteration of a product. It includes style guides, patterns, code components, and other elements that help designers ensure consistency across different parts of a product. By adhering to a Design System, designers can **create more consistent, cohesive products quickly**, as it significantly reduces the guesswork associated with design choices that would otherwise require arbitrary consideration.

Mobile First Design

Mobile First Design is a design strategy that starts by designing for the smallest screen sizes and progressively enhancing the design for larger screens. The approach is based on the idea that **it's easier to scale a design up for a desktop than to simplify a desktop design for mobile**. Mobile First Design ensures that the site is accessible and provides a good user experience on mobile devices, which are increasingly the primary device for internet access.

1. MOBILE

2. TABLET

3. DESKTOP

Responsive Design

Responsive Web Design is an approach to web design that makes web pages render well on a variety of devices and window or screen sizes by changing their layout and content. It aims to provide **optimal viewing and inter-action experience**—easy reading and navigation with a minimum of resizing, panning, and scrolling—**across a wide range of devices**. This approach is essential in a time when mobile devices are commonly used as many user's primary access to the web.

Inclusive Design

Inclusive Design is the process of designing products that are **accessible and usable by as many people as possible, regardless of their age, gender, ability, or status**. It aims to create products that do not exclude anyone and respects diversity among users. Inclusive Design can lead to more accessible products and a larger potential user base.

Responsive design emerged as a reaction to the rapid growth of mobile internet usage. Its history is intertwined with the evolution of mobile devices and their increasingly varied screen sizes and resolutions. The term was first coined by Ethan Marcotte in a 2010 article for "A List Apart."

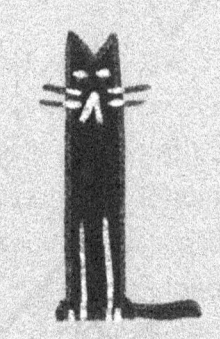

Usability Testing

Usability Testing is a method used in user-centered design that evaluates a product's efficiency, effectiveness, and satisfaction by having **real users interact with it**. The process helps identify usability problems, gather qualitative and quantitative data, and gauge user satisfaction to enable product improvements. From a business perspective, usability testing can help minimize development costs, increase user satisfaction, and boost sales by making the product more user-friendly.

Usability Guidelines

Usability Guidelines are best practices that help design teams create products that are easy to use. They include principles such as **consistency, simplicity, and clear communication**. Following these guidelines can lead to more user-friendly products, higher user satisfaction, and potentially increased adoption rates.

Cognitive Load

Cognitive Load refers to the total **amount of mental effort being used** in the working memory. In design, cognitive load is significant because if a user interface is too complex, it will overload the user's mental capacity, making it difficult for the user to navigate the system. Designers aim to minimize cognitive load, or the "signal to noise" ratio, to make products more user-friendly.

Accessibility

Accessibility in design refers to creating products, devices, services, or environments that are usable by all people, **regardless of their abilities or disabilities**. It ensures that all users, including those with disabilities, have an equal opportunity to interact with and benefit from a product or service. Accessibility can lead to a larger user base and a more inclusive product.

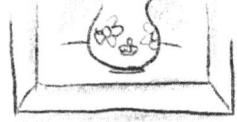

Error Prevention

Error Prevention in design refers to the techniques used to prevent user errors before they occur. This can include **clear instructions, confirmation messages before critical actions, and intuitive design**. Preventing errors can lead to a more user-friendly and efficient product, improving overall user satisfaction.

"The Mother of All Demos" in 1968 by Douglas Engelbart introduced many fundamental elements of modern computing. It showcased windows, hypertext, graphics, efficient navigation and command input, video conferencing, the computer mouse, word processing, dynamic file linking, and a collaborative real-time editor. Engelbart's work is a cornerstone in usability and human-computer interaction.

Strategic & Analytical Approaches

Content Strategy

Content Strategy involves planning, creating, delivering, and managing content. It's about ensuring that the **right content reaches the right user at the right time**. A good content strategy can improve user engagement, increase conversions, and help build a strong brand.

Data-driven Design

Data-driven Design is a design approach that uses data collected from users to inform design decisions. This can include **data on user behavior, preferences, and feedback**. Data-driven Design can lead to more effective and user-friendly products, as design decisions are based on actual user data.

Ethical Design

Ethical Design refers to the practice of designing products that respect the rights and wellbeing of users. It includes **respecting user privacy, being transparent about data usage, and creating inclusive products**. Ethical Design can build trust with users and protect the reputation of a company.

Gamification

Gamification involves applying game design elements in non-game contexts, like a website or app. It's used to motivate **participation, engagement, and loyalty**. Gamification can make an experience more enjoyable, can increase user engagement, and can lead to more user loyalty.

Key Performance Indicators

Key Performance Indicators (KPIs) are measurable values that demonstrate how effectively a company is achieving key business objectives. They help organizations **track their progress and measure their performance against their goals**. Understanding and tracking KPIs can help companies focus on what's important and make better decisions.

A/B Testing

A/B testing is a method of comparing two versions of a webpage or other product to see which one performs better. It involves showing the **two variants to similar visitors at the same time** and comparing which variant drives more conversions. A/B testing can help designers and developers make more informed decisions and create more effective products.

Conversion Rate

Conversion Rate is the percentage of users who **complete a desired action** on a website or app, like making a purchase or signing up for a newsletter. It's a key metric for measuring the success of a website or marketing campaign. A high conversion rate means a product or campaign is effective at getting users to take the desired action.

Click-through Rate (CTR)

Click-through Rate (CTR) is the ratio of users who click on a specific link to the number of **total users who view** a page, email, or advertisement. It's a common metric used to measure the success of an online advertising campaign. A higher CTR indicates a more successful campaign.

Competitor Analysis

Competitor Analysis refers to the process of evaluating the products, strategies, and **strengths and weaknesses of business rivals**. It serves to understand the market landscape, identify opportunities, and inform strategic decisions. In a business context, Competitor Analysis can furnish valuable insights, enabling strategic positioning and better-informed business decisions.

The concept of gamification in education gained traction in the late 20th century. By 1982, academic articles began exploring its potential, following the earlier development of social video games in 1978. The application of game elements like points and badges in learning environments started revolutionizing educational methods.

Consistency

Consistency in design refers to the uniformity of design elements and user experience across a product or platform. This includes **maintaining uniformity in typography, color schemes, imagery, and interface responses to user interaction**. From a business standpoint, design consistency enhances user learnability, usability, and trust, contributing to improved user satisfaction and brand perception.

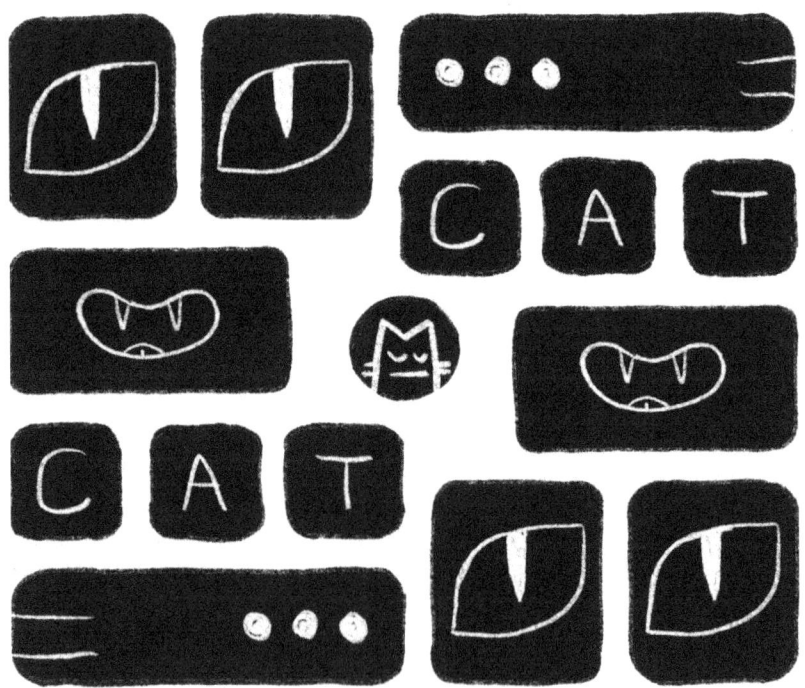

Navigation

Navigation refers to the design elements that **guide users through a product**, like a website or app. Good navigation helps users find the information they need quickly and easily. Effective navigation can improve the user experience and increase user satisfaction.

Above/below the fold

Above the fold refers to the portion of a webpage that's visible without scrolling, while below the fold is the portion that's only visible once you scroll down. These terms come from **newspaper design** and are used in web design to help prioritize the placement of important content. Content above the fold is generally seen as more valuable because it's the first thing users see.

Information Visualization

Information Visualization is the use of visual elements like charts, graphs, and maps to represent information and data. It **makes complex data understandable**, and can help identify patterns, trends, and insights. Effective Information Visualization can make information more accessible, support decision-making, and improve understanding.

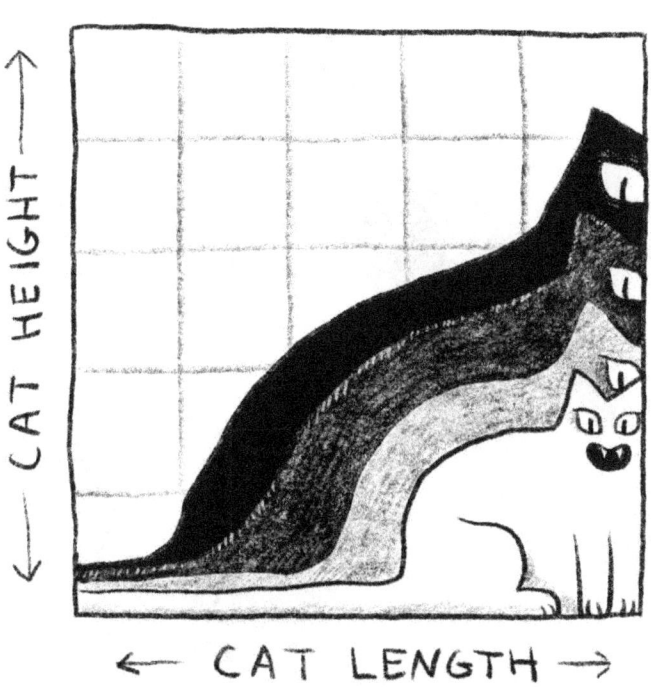

CAT DIMENSIONS

User Engagement

User Engagement refers to the quality of the user's interaction with a product or service. High engagement means the user finds the product **valuable, easy to use, and enjoyable**, leading to repeated use. User Engagement is a key metric for understanding the success of a product or service.

In 1644, Michael Van Langren, a Flemish astronomer, created one of the earliest known visual representations of statistical data. His one-dimensional line graph displayed the varying estimates of the longitudinal difference between Toledo and Rome. This graph was innovative for its time, visually illustrating the discrepancies in data and paving the way for the use of graphical methods in understanding data.

Questions & Answers

Questions & Answers

1. What's with the cat on the cover?

Cats are smart and cats are dumb, a bit like all of us. But cats above all are entertaining. It's true that a visually fun and amusing cat book loaded with facts is an educational one. And that is why you are here, looking at this cat book, to be educated and entertained at the same time. Can it get any better?

2. Why is an illustrated list of terms helpful?

When learning any new topic, a glossary of key terms provides your brain a kind of framework to help organize new incoming information. Adding images with these words further enable your brain to retain this new information. And if you add a touch of humor, the retention is even higher!

3. What is the main benefit of this book?

The main benefit of this book is that you will remember this information and have some fun while doing it.

4. Is "UX for Cats" suitable for complete beginners, or do I need some background knowledge?

"UX for Cats" is perfect for any level of UX knowledge, even if you are a seasoned pro. There is no prerequisite, other than having a sense of humor!

5. What kind of UX principles can I expect to learn from this book?

We cover all of the important and fundamental UX terms you need to get rolling with your next level of learning from a full book or video course. If you read this book, any further UX learning you do will have a ready-made category for that new knowledge to slot right in to.

6. Does the book cover both desktop and mobile design practices?

All the principles covered in the terms listed in this book apply to desktop, tablet, and mobile design processes. They are fully universal. We've got you covered!

7. Who is this book most suitable for?

It was designed for business users and students of any kind that need a fundamental working knowledge of UX in order to be able to understand and contribute to any new design initiatives they might be part of. *Be careful though*: you might find yourself chuckling at the wrong time when you are having powerful sudden recall from the text and images in this memorable book.

8. I'm a business person and will be attending a UX Workshop soon. Will this book help me prepare?

If you have little to no prior UX knowledge, there is no better book on the planet to have in your hands. You will be educated, entertained, and prepared for whatever level of UX participation you are being called to.

9. How does it approach teaching UI concepts compared to other UX books?

Light-hearted illustrations combined with humor and minimal text make it possible to retain the essence of every single term in this book. This is a highly novel approach. We are not sure anything else like this exists!

10. Is this book good for visual learners?

Visual learners with little to no prior UX exposure will be thankful for the day they got this little book, whether they approve of cats or not.

11. Why are there 50 terms and definitions?

49 wasn't enough, and 51 was too much. And 50 struck just the right balance of information with entertainment to be both useful and memorable.

12. Can this book help me with a career in UX design?

This book is a great place to start, as it provides a kind of mental and visual framework to hang all the new facts you are going to accumulate quickly if you make a more formal study of the UX field.

13. Does it discuss the importance of user research and feedback?

Yes, as an integral part of the UX process, you'll find research and feedback topics touch on directly and indirectly in various sections.

14. Is there a focus on the psychological aspects of UX design?

Terms such as "cognitive load" and other psychological factors are also defined and discussed.

15. How does the book make complex UX concepts accessible to younger readers or beginners?

A picture is worth a thousand words! And a funny image is worth a lot more! Having an image to go along with a pithy definition that has an intentionally easy reading level ensures that reader of all ages, but especially younger readers, will have no problem grasping any UX concept presented, no matter how big or small.

16. Does "UX for Cats" mention the latest trends and technologies in UX/UI design?

The core of *UX for Cats* is "laser-pointer" focused on core vocabulary that has already stood the test of time with terms that have not changed much or at all, in some cases, for over twenty years. Trends and technologies change very quickly, but the core work of UX remains stable.

17. What makes this book different from other UX/UI design books?

Taking a humorous and light-hearted approach to a topic that can easily become a little dry and boring is what sets *UX for Cats* apart from other books. If you are a visual learner, this book is great for you. If you are a text-first learner, you'll still get a lot of mileage out of this book especially if you are new to the topic.

18. How comprehensive is the book? Is it more of an overview or in-depth guide?

It was written as an overview and serves as a first point-of-entry to a very broad topic. The 50 terms defined and illustrated are going serve you very well if you have fun with the book and then quickly move on to more in-depth and complete books and course. We suspect you'll be remembering images from *UX for Cats* for years to come, no matter how long you are on your UX journey!

19. Will reading this book alone be enough to start designing user-friendly websites or apps?

This book alone will not get you to a place where you can start designing per se. You are going to want deeper training on fundamental methods of research, wireframing, design principles, and then particular tools as well. But all those things have a core vocabulary that is covered quite well in this book.

20. What if I don't like cats? I'm a dog person!

If you don't like cats, you may have picked up exactly the right book to change your mind. No dogs would cooperate or collaborate on the content or the illustrations, and we aren't sure why. I think cats just are just more cut out for UX and design. Give the book a spin and see if you agree!

21. What role does empathy play in UX design?

Empathy allows designers to understand and anticipate the needs and emotions of users, leading to more considerate and effective designs. More feeling = better design.

22. Can UX design influence customer loyalty?

Absolutely! A well-designed UX can lead to a more satisfying user experience, increasing the likelihood of customers returning and promoting the product.

23. What is "user flow" in UX design?

User flow outlines the path users take within an app or website, highlighting the steps from where they start right through to the final interaction, which helps in designing a smooth and intuitive navigation.

24. How is UX design adapting to the rise of voice interfaces?

UX for voice interfaces focuses on creating conversational, context-aware interactions that feel natural to users, reflecting how people naturally communicate.

25. What is the future of UX design?

The future of UX lies in more personalized experiences, greater emphasis on accessibility, and the integration of AI and machine learning to predict user needs.

26. How does UX deal with data privacy concerns?

UX designers must consider privacy as part of the design process, ensuring that users understand what data is being collected and how it will be used, often through transparent design and clear consent forms. Bad design in this area can get a company in trouble!

27. What are some emerging tools in UX design?

New tools like AI-powered design assistants, advanced prototyping software, and comprehensive user behavior analytics platforms are becoming popular among UX professionals. Expect to see AI take over all the grunt work, and leave strategy as the place designers will have more influence.

28. How important is collaboration in UX design?

Collaboration is crucial in UX design as it involves multiple disciplines. Working together ensures that all aspects of the user experience are cohesive and well-integrated.

29. How does UX design integrate with other disciplines like marketing and product management?

UX design intersects with various other disciplines to enhance product development and user satisfaction. It works closely with marketing to understand user needs and create products that align with market demands, and collaborates with product management to ensure the product meets both user and business goals. Great UX designers are naturally great communicators.

30. What are common UX design mistakes and how can they be avoided?

Common UX design mistakes include failing to understand the user, designing for aesthetics over usability, inconsistent design patterns, and neglecting accessibility. These can be avoided by conducting thorough user research, prioritizing usability, maintaining consistency, and following accessibility guidelines. Assumptions in UX design can be fatal!

31. Can you explain the differences between UX and UI design? Aren't they the same thing?

UX design focuses on the overall experience a user has with a product, considering factors like usability, accessibility, and user satisfaction. UI design, on the other hand, deals specifically with the visual and interactive elements of a product, such as buttons, icons, and layouts. UX encompasses UI, but also extends to a broader range of considerations. So, UI is a part of UX, but not the other way around.

32. What tools and technologies are currently trending in the UX design world?

Some of the trending tools and technologies in UX design include Figma for collaborative interface design, prototyping and wireframing. Additionally, technologies such as voice interfaces, virtual and augmented reality, and AI-powered personalization are becoming increasingly relevant in UX design. Expect AI to fully disrupt UX tools and processes.

33. How is user data protected during UX research and testing?

User data protection is a critical concern in UX research and testing. Best practices include obtaining informed consent from participants, anonymizing data, storing data securely, and adhering to relevant data protection regulations such as GDPR. It's also important to only collect data that is necessary for the specific research or testing purposes.

34. What role does psychology play in UX design?

Psychology plays a significant role in UX design by providing insights into human behavior, perception, and decision-making processes. Understanding psychological principles such as Gestalt laws, cognitive load theory, and social proof can help designers create more intuitive and persuasive user experiences. We touch on some these areas in this book.

35. How can UX design contribute to sustainability and social responsibility?

UX design can contribute to sustainability by promoting digital solutions that reduce environmental impact, such as paperless workflows or energy-saving features. It can also support social responsibility by ensuring products are accessible to diverse user groups, promoting digital inclusion, and considering the ethical implications of design decisions. That said, higher-level concerns like this are far beyond the scope of this little entertaining but enlightening book!

36. What are the career paths available in UX design?

Career paths in UX design include roles such as UX researcher, interaction designer, information architect, UX writer, and UX strategist. With experience, one can progress to positions like senior UX designer, UX team lead, or UX director. There are also opportunities to specialize in specific areas like accessibility, user research, or voice interface design. A close parallel to UX is product owner or manager.

37. How does UX design impact mobile app development?

UX design is crucial for mobile app development as it directly influences user adoption and satisfaction. Key considerations include designing for smaller screens, optimizing navigation for touch interactions, minimizing cognitive load, and ensuring fast load times. Good UX design can make the difference between a successful app and one that struggles to retain users.

38. What is the importance of color psychology in UX design?

Color psychology plays a significant role in UX design by influencing user emotions, perceptions, and behaviors. Different colors can evoke specific feelings and associations, such as trust, excitement, or calmness. Choosing the right color scheme can enhance usability, guide attention, and reinforce brand identity. However, you'll notice little book is black and white. This is very much in purpose, to keep the information, well, "black and white" and not distract from the core meaning of each term, and each illustration that goes with it. The decision to go black and white with this book, therefore, was a truly psychological one!

39. What exactly is a user interview?

Conducting effective user interviews involves preparing a clear discussion guide, recruiting representative participants, and creating a comfortable environment for open dialogue. Key techniques include asking open-ended questions, practicing active listening, and avoiding leading or biased questions. The goal is to gain deep insights into user needs, behaviors, and pain points by making the whole process as relaxed and natural as possible for the best result.

40. What are some best practices for creating a user-friendly e-commerce website?

Best practices for creating a user-friendly e-commerce website include having a clear and intuitive navigation structure, providing high-quality product images and descriptions, offering easy search and filtering options, and streamlining the checkout process. It's also important to ensure the site is mobile-friendly, loads quickly, and provides helpful customer support features. Bad design = bad sales.

Illustration & Memorization Gallery

Test Your Visual Memory

Welcome to the gallery section of UX for Cats! You've pounced this far, and now it's time to put your knowledge to the test.

Here, you'll find images separated from their definitions, giving you the perfect opportunity to practice recalling what you've learned. This method reinforces memory by challenging you to remember terms through visual cues alone.

Dive in, have fun, and embrace the playful challenge. Good luck, and may your memory be as sharp as cats claws!

Foundational Concepts:
User Experience

Foundational Concepts:
User Interface
Design Principles

Foundational Concepts:
User-Centered Design

Foundational Concepts:
Design Thinking

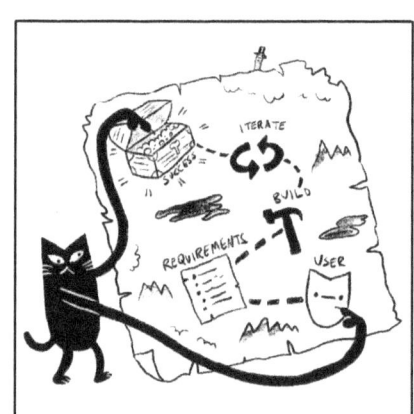

Foundational Concepts:
Information Architecture

Research & User Analysis:
User Persona

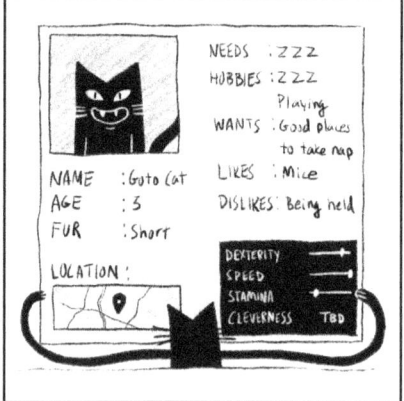

Research & User Analysis:
User Journey

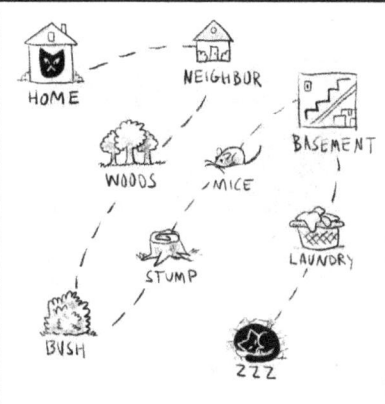

Research & User Analysis:
User Flow

Research & User Analysis:
User Feedback

Research & User Analysis:
User Interviews

Research & User Analysis:
User Surveys

Research & User Analysis:
User Stories

Research & User Analysis:
Task Analysis

Research & User Analysis:
User Research

Research & User Analysis:
Empathy

Design & Prototyping Processes
Prototyping

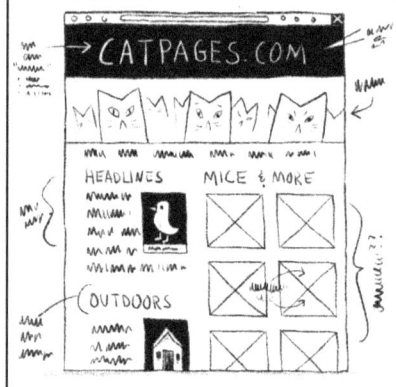

Design & Prototyping Processes
Wireframing

Design & Prototyping Processes
Rapid Prototyping

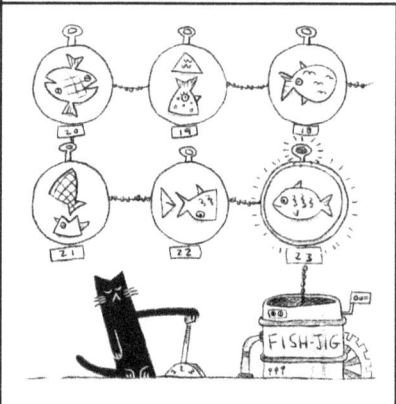

Design & Prototyping Processes
Iterative Design

Design & Prototyping Processes
Lean UX

Visual & Interaction Design
Visual Design

Visual & Interaction Design
Typography

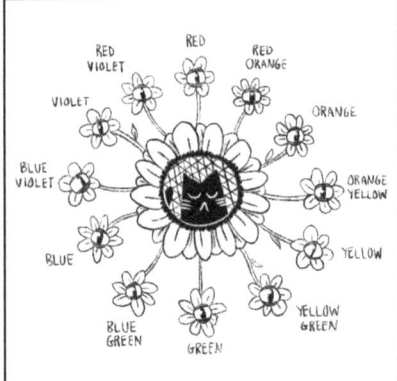

Visual & Interaction Design
Color Theory

Visual & Interaction Design
Visual Hierarchy

Visual & Interaction Design
Fidelity

Visual & Interaction Design
Interaction Design (IxD)

Development Methodologies
Agile Development

Development Methodologies
Design System

Development Methodologies
Mobile First Design

Development Methodologies
Responsive Design

Development Methodologies
Inclusive Design

Usability & Accessibility
Usability Testing

Usability & Accessibility
Usability Guidelines

Usability & Accessibility
Cognitive Load

Usability & Accessibility
Accessibility

Usability & Accessibility
Error Prevention

Strategic & Analytical Approaches
Content Strategy

Strategic & Analytical Approaches
Data-driven Design

Strategic & Analytical Approaches
Ethical Design

Strategic & Analytical Approaches
Gamification

Strategic & Analytical Approaches
Key Performance Indicators

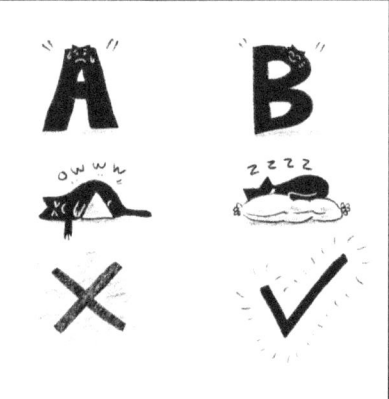

Strategic & Analytical Approaches
A/B Testing

Strategic & Analytical Approaches
Conversion Rate

Strategic & Analytical Approaches
Click-through Rate (CTR)

Strategic & Analytical Approaches
Competitor Analysis

Additional Concepts
Consistency

Additional Concepts
Navigation

Additional Concepts
Above/below the fold

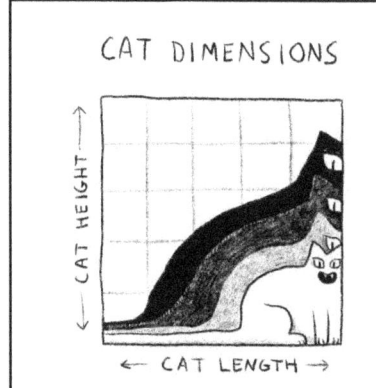

Additional Concepts
Information Visualization

Additional Concepts
User Engagement

What to Read Next

Beginner

These books provide a solid introduction to fundamental UX design principles, methodologies, and practical approaches, suitable for those new to the field.

"Don't Make Me Think" by Steve Krug
A must-read on web usability that is highly accessible.

"The Design of Everyday Things" by Donald A. Norman
Introduces the key concepts of user-centered design in everyday context.

"100 Things Every Designer Needs to Know About People" by Susan Weinschenk
Offers insights into human behavior crucial for designing intuitive interfaces.

"The Elements of User Experience" by Jesse James Garrett
Outlines the fundamental components that make up user experience design.

"Universal Principles of Design" by William Lidwell, Kritina Holden, and Jill Butler
A broad introduction to essential design principles.

"UX for Beginners: A Crash Course in 100 Short Lessons" by Joel Marsh
Breaks down the core concepts of UX into manageable lessons.

"Rocket Surgery Made Easy" by Steve Krug
Guides beginners through simple techniques for usability testing.

"Sketching User Experiences: Getting the Design Right and the Right Design" by Bill Buxton
Introduces the importance of sketching in the UX process.

"About Face: The Essentials of Interaction Design" by Cooper, Reimann, Cronin, and Noessel
Comprehensive guide on interaction design principles.

**"Information Architecture: For the Web and Beyond"
by Louis Rosenfeld, Peter Morville, and Jorge Arango**
*Key concepts for structuring and organizing web
information.*

Intermediate

These books delve deeper into specific aspects of UX design, suitable for those who already have a basic understanding and are looking to expand their knowledge and skills.

**"Hooked: How to Build Habit-Forming Products" by
Nir Eyal**
*Explores advanced product design strategies that encourage
user engagement and habit formation.*

"Designing with the Mind in Mind" by Jeff Johnson
*Provides an in-depth look at how understanding cognitive
functions can improve interface design.*

**"A Project Guide to UX Design" by Russ Unger and
Carolyn Chandler**
*Offers insights into managing UX projects, perfect for those
stepping into leadership roles or managing larger projects.*

**"Emotional Design: Why We Love (or Hate) Everyday
Things" by Don Norman**
*Dives into the affective aspects of design, explaining why
emotions play a significant role in UX.*

**"The User Experience Team of One: A Research and
Design Survival Guide" by Leah Buley**
*A great resource for UX practitioners who need to implement
UX strategies and processes on their own.*

Index

A

A/B Testing 69, 108

Above/below the fold 78, 110

Accessibility 59, 106

Additional Concepts 75

Agile Development 48, 103

AI 88

B

beginners 84, 86

business person 85

C

career 86, 90

cat 84

Click-through Rate (CTR) 71, 109

Cognitive Load 58, 106

color psychology 91

Color Theory 42

Competitor Analysis 72, 109

Consistency 76, 110

Content Strategy 64, 107

Conversion Rate 70, 109

customer loyalty 87

D

data privacy 88

Data-driven Design 65, 107

Design & Prototyping Processes 31

design mistakes 89

Design Principles 95

Design System 49, 104

Design Thinking 15, 96

desktop 85

Development Methodologies 47

dog 87

E

emerging tools 88

Empathy 29, 87, 99

Error Prevention 60, 106

Ethical Design 66, 107

F

Fidelity 44, 103

Foundational Concepts 11

future 88

G

Gamification 67, 108

I

Inclusive Design 52, 105

Information Architecture
16, 96

Information Visualization
79, 111

Interaction Design (IxD)
45, 103

Iterative Design 35, 101

K

Key Performance
Indicators 108

Key Performance Indicators
68

L

Lean UX 36, 101

M

memory 94

mobile 85, 91

Mobile First Design 50,
104

N

Navigation 77, 110

P

Prototyping 32, 100

psychological aspects 86

psychology 90

R

Rapid Prototyping 34, 100

Research & User Analysis
19

Responsive Design 51,
104

S

social responsibility 90

Strategic & Analytical
Approaches 63

sustainability 90

T

Task Analysis 27, 99

trends 86

Typography 41, 102

U

Usability & Accessibility 55

Usability Guidelines 57, 105

Usability Testing 56, 105

User Engagement 80, 111

User Experience 12–13

User Feedback 23, 97

User Flow 22, 88, 97

User Interface
Design Principles 13

user interview 91

User Interviews 24, 98

User Journey 21, 97

User Persona 20, 96

User Research 28, 99

User Stories 26, 98

User Surveys 25, 98

User-Centered Design 14, 95

UX principles 84

V

Visual & Interaction Design 39

Visual Design 40

Visual Hierarchy 43

visual learners 85

Visual Memory 94

voice 88

W

Wireframing 33, 100

Feedback

I ~~want~~ need to hear from you!

Join our **newsletter**!
Send a **comment**!
Send a *testimony*
Submit a **review**!
…or just say "meow":

feedback@gotocats.com

Other Titles

The Big Book of Font Combinations

The Preposterously Huge Book of
Google Font Combinations

About the Artist & Author

Douglas Bonneville is an artist, designer, and developer. His background is in fine art, and he brings that sensibility to his illustration, graphic design, and code projects.

He has written about design, UX, art, artificial intelligence and more at his blog BonFX, and has produced books of his own and for others.

As a child very interested in art, he was deeply influenced by the illustrations and limericks in the "Book of Nonsense", by Edward Lear, and gravitated towards pen and ink based styles. Other influences were Gary Larson, Edward Gorey, Dr. Seuss, and Richard Scarry. An obscure book from the 1980's called "The Big Book of Amazing Facts" by Malvina Vogel & Mel Mann was deeply influential on his love of crafting certain kinds of books.

Douglas lives in Rhode Island with his wife (and their cats). Find him on X at @dbonneville and say "meow".

www.ingramcontent.com/pod-product-compliance
Lightning Source LLC
Chambersburg PA
CBHW021146070326
40689CB00044B/1138